T5-ACQ-525

INCREDIBLE INVENTORS

GRACE HOPPER AND THE COMPUTER

BY BENJAMIN PROUDFIT

Gareth Stevens
PUBLISHING

Please visit our website, www.garethstevens.com. For a free color catalog of all our high-quality books, call toll free 1-800-542-2595 or fax 1-877-542-2596.

Library of Congress Cataloging-in-Publication Data
Names: Proudfit, Benjamin, author.
Title: Grace Hopper and the computer / Benjamin Proudfit.
Description: New York : Gareth Stevens Publishing, [2023] | Series: Incredible inventors | Includes index.
Identifiers: LCCN 2021036575 (print) | LCCN 2021036576 (ebook) | ISBN 9781538276587 (set) | ISBN 9781538276594 (library binding) | ISBN 9781538276570 (paperback) | ISBN 9781538276600 (ebook)
Subjects: LCSH: Hopper, Grace Murray–Juvenile literature. | Women computer engineers–United States–Biography–Juvenile literature. | Computer engineers–United States–Biography–Juvenile literature. | Women admirals–United States–Biography–Juvenile literature. | Admirals–United States–Biography–Juvenile literature.
Classification: LCC QA76.2.H67 P76 2023 (print) | LCC QA76.2.H67 (ebook) | DDC 621.39092 [B]–dc23
LC record available at https://lccn.loc.gov/2021036575
LC ebook record available at https://lccn.loc.gov/2021036576

Published in 2023 by
Gareth Stevens Publishing
111 East 14th Street, Suite 349
New York, NY 10003

Copyright © 2023 Gareth Stevens Publishing

Designer: Rachel Rising
Editor: Kristen Nelson

Photo credits: Cover, p. 1 National Commission on Military, National, and Public Service. (9/19/2017 - 9/18/2020). Retrieved from the Digital Public Library of America; pp. 3-24 Zavadskyi Ihor/Shutterstock.com; pp. 3-24 Lukasz Szwaj/Shutterstock.com; p. 5 Grace Murray Hopper Collection, 1944-1965, Archives Center, National Museum of American History; p. 7 Winston Tan/Shutterstock.com; p. 9 https://commons.wikimedia.org/wiki/File:Harvard_Mark_I_Computer_-_Input-Output_Details.jpg; p. 11 https://commons.wikimedia.org/wiki/File:Grace_Hopper_and_UNIVAC.jpg; p. 13 https://commons.wikimedia.org/wiki/File:Grace_Murray_Hopper,_in_her_office_in_Washington_DC,_1978,_%C2%A9Lynn_Gilbert.jpg; p. 15 Trismegist san/Shutterstock.com; p. 17 Department of Defense. American Forces Information Service. Defense Visual Information Center. 1994. Retrieved from the Digital Public Library of America; p. 19 Department of Defense. American Forces Information Service. Defense Visual Information Center. 1994. Retrieved from the Digital Public Library of America; p. 20 https://commons.wikimedia.org/wiki/File:Presidential_Medal_of_Freedom.svg.

All rights reserved. No part of this book may be reproduced in any form without permission in writing from the publisher, except by a reviewer.

Printed in the United States of America

Some of the images in this book illustrate individuals who are models. The depictions do not imply actual situations or events.

CPSIA compliance information: Batch #CSGS23: For further information contact Gareth Stevens, New York, New York at 1-800-542-2595.

Find us on

CONTENTS

From the Beginning 4

In the Navy 8

Programming 10

Using Words 12

Navy Life 16

A Teacher Too 18

Honored 20

Glossary 22

For More Information 23

Index 24

Boldface words appear in the glossary.

From the Beginning

Grace Hopper played a big part in the invention of the computer. She was born Grace Murray on December 9, 1906. She lived in New York City with her family. Grace always liked to find out how things worked. She took apart many alarm clocks!

5

Grace went to Vassar College and finished there in 1928. She went on to Yale University to study math. In 1930, she got married and became Grace Hopper. She earned her master's **degree** that year and a **doctorate** four years later.

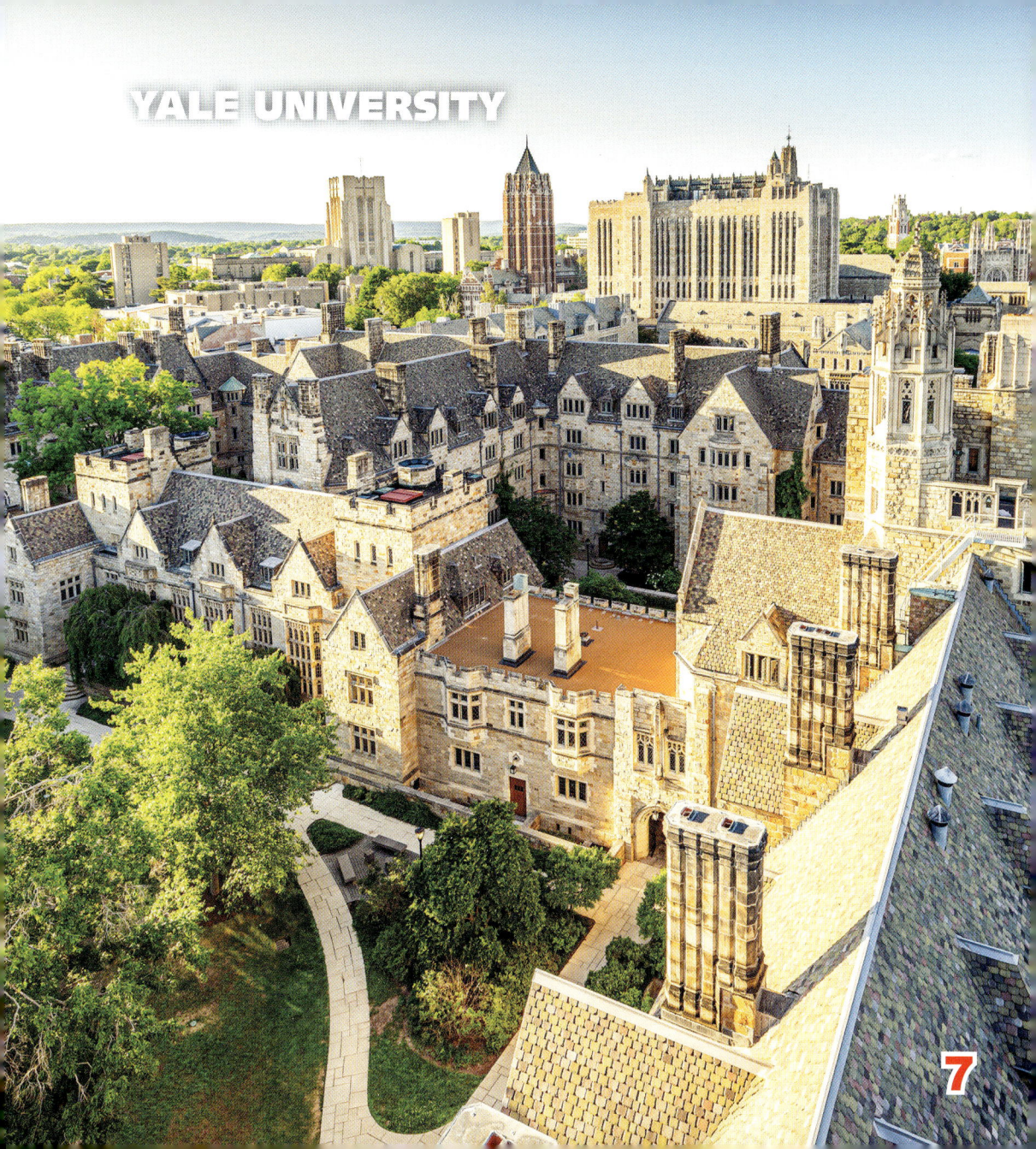

YALE UNIVERSITY

7

In the Navy

The United States entered **World War II** in 1941. Grace wanted to help. She joined the U.S. Naval **Reserve** in 1943. She was sent to work on an early computer with a team at Harvard University. The computer was called MARK I.

MARK I

Programming

On the navy's computer teams, Grace became one of the first **programmers**. In 1949, after the war, she took those skills to a company. She worked on other important early computers, including UNIVAC I and II, during the early 1950s. She made the first **compiler**!

UNIVAC I

11

Using Words

Up until then, computer programs were written using **symbols**. Grace thought more people could use computers if programs used words. Her idea was new! She was told it wouldn't work. She kept working on it anyway.

13

In the mid-1950s, Grace's team began running a programing language called FLOW-MATIC. It used English word commands. Grace then worked on coming up with a common computer language for use in business. It was called COBOL. It was a big success!

15

Navy Life

Grace served in the U.S. Navy until she was 79 years old. She worked on writing the navy's computer languages and programs. Some people called her "Amazing Grace!" When she left the navy, she held the high **rank** of rear admiral.

17

A Teacher Too

Grace was a teacher as well as a programmer. She taught at Vassar College and others. She also taught community classes about computers. Grace said of all the things she did in her life, teaching brought her the most happiness.

19

Honored

On January 1, 1992, Grace Hopper died. In 1996, the U.S. Navy named a ship—the USS *Hopper*—for her. In 2016, she was awarded the Presidential Medal of Freedom. Her contributions to computer science are still honored today.

PRESIDENTIAL MEDAL OF FREEDOM

TIMELINE

December 9, 1906 — Grace Murray is born in New York City.

1928 — She finishes school at Vassar College.

1930 — She gets married and becomes Grace Hopper. She earns a master's degree.

1934 — She earns her doctorate from Yale University.

1943 — She joins the U.S. Naval Reserve.

1950s — Grace is part of the teams that work on MARK I, UNIVAC, FLOW-MATIC, and COBOL.

1986 — She leaves the U.S. Navy at age 79.

January 1, 1992 — Grace Hopper dies.

1996 — A naval ship is named for her.

2016 — She is awarded the Presidential Medal of Freedom.

GLOSSARY

compiler: a computer program that turns a set of directions written in symbolic language into computer language so the directions can be followed

contribution: something that is done to cause something to happen

degree: a document and title earned by finishing a series of classes at a college or university

doctorate: the highest degree offered by a university. It requires many years of study.

programmer: a person who makes and tests programs for computers

rank: an official standing

reserve: referring to soldiers who are not part of a country's main forces but may be called to active duty in times of need

symbol: a picture, shape, or object that stands for something else

World War II: a war fought from 1939 to 1945 that involved countries around the world

FOR MORE INFORMATION

BOOKS

Holub, Joan, and Daniel Roode. *This Little Scientist: A Discovery Primary.* New York, NY: Scholastic, 2018.

Loewen, Nancy. *Grace Hopper: The Woman Behind Computer Programming.* North Mankato, MN: Pepple, a Capstone imprint, 2020.

WEBSITES

Fun Computer History Facts for Kids
easyscienceforkids.com/all-about-computers
Read about the history of computers here.

This Is Grace
www.timeforkids.com/g56/this-is-grace-hopper
Find out even more about the life of Grace Hopper in this article.

Publisher's note to educators and parents: Our editors have carefully reviewed these websites to ensure that they are suitable for students. Many websites change frequently, however, and we cannot guarantee that a site's future contents will continue to meet our high standards of quality and educational value. Be advised that students should be closely supervised whenever they access the internet.

INDEX

COBOL 14, 21

family 4

FLOW-MATIC 14, 21

Harvard University 8

MARK I 8, 21

New York City 4, 21

Presidential Medal of Freedom 20, 21

UNIVAC (I and II) 10, 21

U.S. Naval Reserve 8, 21

U.S. Navy 10, 16, 21

USS *Hopper* 20

Vassar College 6, 18, 21

Yale University 6, 21